Robert McKinley Ormsby

Darwin

Or God in Nature

Robert McKinley Ormsby

Darwin
Or God in Nature

ISBN/EAN: 9783743432871

Manufactured in Europe, USA, Canada, Australia, Japa

Cover: Foto ©Lupo / pixelio.de

Manufactured and distributed by brebook publishing software
(www.brebook.com)

Robert McKinley Ormsby

Darwin

DARWIN;

OR,

God in Nature.

" I am ready to allow, * * * that the ultimate cause of all motion is immaterial—that is, God." *Zoonomia*, vol. I, § 14.—ERASMUS DARWIN, 1794.

BY ROBERT McK. ORMSBY.

" Si quid novisti rectius istis,
Candidus imperti: sed non, his utere mecum."

SECOND EDITION.

New York:

MASONIC PUBLISHING AND FURNISHING COMPANY,

Barker, du Laurans & Durham,

729 BROADWAY.

1878.

P. F. McBREEN, Printer and Electrotyper,
14 & 16 Ann Street, New York.

PREFACE.

FIRST EDITION.

MR. HUXLEY recently gave three lectures in New York City, on the horse, showing that at a prior geologic period, he had four toes, and arguing that he was developed from a five toed, or hoofed mammal. In his lectures, he assumed that the world had a beginning. In speaking of " the hypothesis of the eternity of this state of things in which we now are," he remarked that, " whether true or false, it is not capable of verification by evidences;" and there dropped it. This was a masterly *coup de logique!* The permanence of the laws and order of nature is a settled presumption, and it is for him who assumes the contrary to produce his proof. The burden of evidence being on the learned professor, he will do well to advance his facts; and assumed facts will not answer. He has only one authority; that is the Bible. Was it on this account that he so tenderly spared Moses, in his lectures, at the expense of the poet Milton? I hope not, for in questions of science, we must not believe Moses nor the prophets.

The learned professor's New York lectures have provoked the evolution of this poem, and I have only to beseech him not to treat it as he did the horse, by too closely scanning its feet; because, if he does, I feel certain he will make the feathers fly, and claim they indicate a fowl origin.

Chester Hill, Mt. Vernon,
*   New York,*
November, 1876.

PREFACE.

SECOND EDITION.

Almost every scientist, philosopher, geologist and biologist in the world has favorite theories of cosmogony, geology and biology, and will be apt to receive with coldness those of an adverse character. But Nature is a kind mother ; she has a large empire, with wide domains, wherein there seems a place for everything—this author's little poem amongst the rest. The first edition was quickly absorbed, taken by candid and impartial readers, from many of whom such cordial commendations have been received by the author as to encourage the issue of another. This edition is accompanied by notes, and is respectfully submitted to the judgment of the public.

Chester Hill,
Mt. Vernon. New York,
Oct. 30, 1877.

PROEM.

A New Theory in regard to the Order of Geological Formations. No Glacial Period ever existed.

It is sometimes assumed that the earth was evolved, with the other planets and the sun, from nebula. This nebula, astronomers have said, was a heated, gaseous substance. The earth was incandescent at first, and cooled slowly. The crust of the earth is the result of the cooling process, thought to be still in progress, the central part being a heated, fused mass.

But, it is submitted, that there is not sufficient, if any, evidence to show: first, that the solar system was ever in a gaseous state; second, that the earth was ever incandescent; third, that the centre of the earth is in a fused condition; and, fourth, that there was ever a first (azoic) crust of the earth.

As to the first point, science reveals but one way of transforming solids into gas, and that is by heat. The mother nebula must have existed in space from eternity, or been created from solids. That nebula is changed into suns and planets, and suns and planets into nebula, is only possible by laws of which we have no knowledge. There is not heat enough in the solar regions to resolve the planets into gas. After our system lost its heat, which departed into space, as we are told, our planets appeared, there not being heat enough left to keep them in vapor. A very astonishing proposition !

As to the second point, it may be asked how geologists know that the earth was ever incandescent; and, thirdly, that the centre is warmer than the surface? Here, inferences are drawn from volcanic phenomena; from the existence of plutonic below stratified rock; and from the fact that as we penetrate into the earth the temperature rises at the rate of one degree for every forty-five feet. According to this last item, calling the temperature at the surface fifty degrees, at the depth of thirty miles a heat would be encountered sufficient to fuse most known substances. This would place our existence upon a very thin crust of a globe of fire! According to this view, we have bottled up in the interior of our globe a large specimen, or rather, remnant, of the material out of which the shining worlds around us were made!

To such brilliant conjectures, reason, enlightened by science and experience, offers the suggestion that the heat below the surface of the earth is at but a shallow depth from the surface, being merely at the base of the oxydized part of the earth. Between this point and the centre, or central cavity, if one, the distance is very great. It would not be unreasonable to suppose that the forms of all bodies of the universe are in accordance with the laws displayed in the works of nature; and no one of nature's laws is more striking than that which so shapes or forms objects as to give to the least amount of matter the greatest strength and usefulness. Analogical reasoning would not infer a globe solid to the centre. All the changes of the earth which occur from age to age are limited to the surface, and are caused by water and heat. But the heat is generated at certain depths by chemical action. All changes which matter undergoes below the surface of the earth are attended, of course, with the disengagement of heat; and such changes cannot occur at a point not reached by moisture. In some localities the changes are more active than at others, as indicated by the temperature. Increase of temperature is not always experienced in penetrating the crust of the earth, nor is the degree of increase in all places the same.

The fourth point is, that a crust, being a series of plutonic rock, was formed over the surface of an incandescent globe from the effects of the cold of celestial space, just as from contact with cold air a crust is formed over a stream of lava.

But, in regard to the primary formation, the opinions of geologists vary. It has been claimed that the fragments of the strata of an antecedent formation have been found included in this; and the evidences that life was, during its period, extant, are being discovered by scientific gentlemen. It was of very great thickness, and is said now to rest upon granite; but we may perhaps suppose that when its deposition commenced it was upon a regularly stratified formation, then existing. A careful examination of the appearance and character of the various strata of the metamorphic rock will lead many to believe that its materials are derived from disintegrated, stratified rock. And the axes of the New England primary formation are about twenty miles apart. As this very great depth of deposit was probably upon an older deposit of equal thickness, or, if tilted, of vastly greater depth, the latter must have found itself, as the superimposed deposit increased, constantly approaching the regions of intense heat. At such depth every metal and metaloid of a very volatile character would be dissipated from the ancient formation, and what remained would be fused granite, ready, at the period of upheavel, to be extruded in mountains, like the granite mountains of New Hampshire, and other granite mountains where tilted primary strata are found. Thus every strata, as it works its way downwards, will, sooner or later, come again to the surface in shape of granite, to re-enter in the career of transformation. In volcanos the more and less volatile metals may be mingled, or the material escaping be of the more volatile character.

Geologists say they find a period when life first existed on our globe. The primary fossiliferous rock in England is located in the region anciently called the Silures. There the remains of low organic forms are found. At that supposed period, the trilobite, a sort of bug, seems to have held a high place amongst organic beings, and to have existed prior to the age of plants. The Silurian is a vast system, and would be supposed to have drawn largely from the alluvium of some tributary continent; but it was built up at a period when the earth is not credited with continents, nor alluviums.

Different latitudes give different faunas; and England,

in her formations, appears to have the faunas of every zone. Her tropical, if not her arctic and antarctic faunas have been several times repeated; and, still, it has not been suggested that she has seen other latitudes. The earth cannot be stationary. The inclination of the equatorial plane to the plane of the ecliptic probably never changes, so that the position of the cold or polar regions in the heavens is always the same; but under the icy masses the earth is continually revolving, having perhaps no fixed axis in the equatorial regions, but moving with a spiral motion. The Atlantic basin, with northern Europe and America, are but fresh from the arctic ice, as indicated by the freshness of the striæ upon the polished ledges. The direction of these striæ is to the receding polar masses. In this manner tropic lands go to the northern and southern temperate zones; to the arctic and antarctic poles; pass under the icy polar mountains, and keep on going around and round forever. Thus tropic fossils are conveyed around the earth upon her bosom; and thus we find Asiatic forests and coal beds, if not fossil alluvium bones, emerging from under the polar ice. Those great mills of the gods, the polar icebergs, are, in a great measure, our soil makers, giving us our drift, our alluvium, and spreading over the earth an infinite quantity of rounded, polished pebbles, and scattering, as the tracks of floating icebergs, large boulders of drift rock. This continual change of the earth's polar axis is too slow and gradual to be perceived only after the lapse of ages. Astronomers may, and perhaps do, detect it.

They tell us that the celestial latitudes, in certain situations, increase, and in the opposite, diminish, indicating a change of the earth's inclination at the rate of forty-eight seconds in a century. If the change of inclination be only apparent from the change of the place of the spectator, and this apparent change be owing to the supposed revolution of the earth, the rate of progress may be easily calculated. An entire revolution would require 2,700,000 years; and our native land would have emerged from the arctic ice about 350,000 years ago. This is assuming, however, that the terrestrial latitudes keep pace with the celestial. The celestial latitudes being based on the ecliptic may change faster than the terrestrial which are based on the equator. The equatorial line shares in the change which is mixed with the precession phenomenon; but this movement should not be misapprehended. We may consider the equatorial line and the lines of latitude, iron rings about the earth. They never move. The position of all, as well as of the polar glaciers, forever remain stationary. The movement, if any, is of the earth beneath them.

There is nothing in this movement that disturbs the order of superposition of strata as noted by geologists.

Faunas and floras are mainly the creatures of latitude; and formations are always determined by the faunas and floras they contain. "A group of strata," says Agassiz, "extending over a certain geographical extent, all of which contain some fossils in common, no matter what may be the chemical character of the rock, whether it be limestone, sand or clay, is termed a geological formation." As these formations are thus identified by their fossils, if we must seek a suitable place for their origin, we must select latitudes which are the homes of the faunas and floras from which these fossils were derived. The Carboniferous is a tropical formation, and it would not be rash to suppose it to have been laid down at or on the equator. But while that formation was being deposited, other formations, in latitudes where there were other faunas north or south of it, were also being laid down. The Trias, the Oolite and the Cretaceous, which all partake of the tropical character, may have been deposited contemporaneously with the Carboniferous, upon one side of it, either north or south, while the Devonian and Silurians were deposited upon the other. The formations may thus all be contemporaneous; every one known to geology being made at the same time, side by side, from the Arctic to the Antarctic Oceans. A geologic age will complete each; but we must reflect that at the end of that, which we will call the first age, the earth has changed its position in relation to the latitudes. That part of the earth which was on the equator, and had received its Carboniferous burden, has moved towards the temperate zone. The Devonian being next to, and following, will, at the commencement of the second age, be on the equator, so that the Carboniferous deposit of this age will be over the Devonian. As the climate determines the fauna, and the fauna the formation, no two ages will deposit a particular formation in the same place upon the earth, although always in the same latitude.

As we conjecture that Europe and America are passing south, we must suppose that Asia is moving north. The Silurian is said to have a southern antarctic cast, so we will locate the place of its origin somewhere on the southern antipodes, or on the side of the globe opposite to us. On the equator, of course, we must place the Carboniferous; the Devonian next south; then the Silurians and Cambrian further south. On the north, next to the Carboniferous, we place, first, the Trias; next, the Oolite; next, the Cretaceous; next, the Tertiary. We may suppose these formations, repeated six or eight times, making as many ages. The order of superposition will not vary from that taught by geology. It may thus be illustrated.

A diagram of the deposits of seven ages, during which this section of the earth's surface is continually moving north fast enough to displace, during an age, each of its formations.

AGES.

```
7th...........................................Tert.,   &c.
6th.......................................Tert.,  Cret.,  &c.
5th....................................Tert.,  Cret.,  Trias,  &c.
4th................Tert.,   Cret.,  Trias,  Oolite, &c.
3d.............Tert., Cret.,  Trias, Oolite, Carb., &c.
2d.......Tert., Cret., Trias,  Oolite, Carb., Dev.,   &c.
1st..Tert., Cret., Trias, Oolite., Carb., Dev.,  Sil.
```

N——————————————————————————————————S

The above line—a longitudinal section of the earth.

By this diagram it is seen that during each geological age, in which all the formations may be deposited, the earth moves north the number of degrees of latitude covered by each formation. This will leave for the second age a change of position of the formations, as the faunas which designate the formations lag behind, sticking to the latitudes, and **not** following the migrating deposits of each age. Geology abounds in evidence of this supposed motion of the earth. Lyell says (and everybody can see it), that in the fossils of the transition rocks a tropical and humid atmosphere is indicated. The Carboniferous formation he pronounces the gift of a tropical climate. The vegetation, says he, during parts of the period between the Lias and the Chalk, appears to have approached that of the large islands of the equatorial zone. And between the uppermost member of the Secondary, and the oldest of the Tertiary, he tells us the fossils indicate the prevalence of a very hot climate. And Agassiz remarks that there had been an ice age, or, as he expresses it, a reduction of temperature before each great geologic age. And later geologists claim as many as nine ice ages prior to our last glacial period, as indicated by the strata from the Silurian up to the Pliocene.

However, it is well known that other things, besides latitudes, determine faunas and floras. Difference of longtitude sometimes has effect, as well as depth of ocean, height of country, and the substances held in solution in waters.

Furthermore, it would not be possible, except in theory, for all the formations to be deposited in any one age. We can only say that each formation must have its appropriate latitude or home, but other things must occur to insure its construction. To have a deposit of a regular series of formations, side by side, from the north to the south poles, in the same age, no one could expect, as such a result would require an ocean from pole to pole, with suitable continental elevations from which such formations may be made. And geologists find these breaks in the order, and are puzzled over them. Although the series is not anywhere full and perfect, the order of superposition must, by the movement supposed, be always the same. During the present geologic age there may be no Carboniferous formation in progress, or there may be no Cretaceous; but, when these do find periods of birth, there can be but one latitude in which each may come into existence. Lyell notices the breaks in the series of superimposed formations, which are common in the Secondary and Tertiary, but more frequent in the latter, and says they are yet to be accounted for. They occur everywhere. In no one spot on the earth have we found all the formations present, and in many places but one, and in some places none, except the drift above the primary.

Then, the gradual change of latitude of the formations should occasion some mingling of faunas of adjoining formations; and the more or less gradual change of faunas and floras is remarked by geologists. There are some changes of fauna that seem abrupt; a thing not at all unaccountable, according to the theory here broached, as it never occurs without a break in the order of formations.

There are two systems of geological formation to be regarded by the paleontologist. What is called the Secondary system, ending with the Chalk formation, should be regarded as oceanic. Sub-continental features in the Secondary do occasionally occur, but comparatively not to great extent. In no member of this series should we look for fossils of mammalia, or land quadrupeds of the mammalian type. Should a continent now be raised up out of the Atlantic Ocean, we should not find in the fractured strata of its main bed anything but the lower species of plants, fish and mollusca, with no vestige of the continental life of the present or past ages. It would not be impossible, of course, for a fragment of some mammal of the Rocky Mountains, by means of the Mississippi River and the Gulf Stream, to find its way into the bed of the Atlantic; but, such transportation, though not impossible, should not be expected. And the fact that none of the oceanic deposits contains the fossils of land animals, with rare exceptions, is far from raising a presumption that those deposits occurred in an age when no land animals lived.

The Trias, Oolite and Wealden, of the upper Secondary, have the sub-continental features alluded to. These are tropical and sub-tropical. They include remains of Ophidian, Saurian and Chelonian animals and other inhabitants of equatorial latitudes. In speaking of the Wealden, Mr.

Lyell remarks :—"In no part has any fragment of the "skeleton of a mammiferous quadruped been obtained. "The strata of the Wealden, with one exception, present "such characters as we might look for in the deposits of "the deltas now forming at the mouths of large rivers in "tropical climates."

It may be asked why we have so many formations of a tropical character; but the reader must bear in mind that if the earth has the revolution here claimed, it must have a slowly varying axis whose polar regions cannot leave the tropics only after an immense number of revolutions, and without a spiral motion never. We get these tropical formations by means of the spiral motion, before alluded to.

The Tertiary has our principal record of the existence of mammalian quadrupeds. The Tertiary, as contra-distinguished from oceanic formations, is substantially continental in its origin. When we call the Tertiary a continental formation, we add the word sub-continental, as the system embraces not only the deposits in continental seas and lakes, but also coast deposits in surrounding oceans. It is not asserted that mammalian life originated in our Tertiary formation. It was, undoubtedly, derived from the continents from which our Tertiary was formed—continents now mostly submerged in our present oceans.

Geologists arrange the Tertiary formations as follows : the lowest (oldest), Eocene; next, Miocene; then Pliocenes (old and new); and lastly, newest of all, Recent and Alluvium. We are told that man and most of the domestic animals, being found in the alluvium, must have had a recent origin. But the truth is, the alluvium is, or may be, older than Tertiary formations, and it may therefore hold the remains of earlier land animals. The theory of this paper would make the alluvium embrace all detrital soils not in rocky strata, including the drift. It is only when there are Mediterranean seas, lakes and river deltas off the coast, that Tertiary strata can be deposited, during which time portions of the old detrital soils of the continents are being wasted, and new strata (called Tertiary) deposited in far away corners. Compared with a great continent, the Tertiary formations of any particular age embrace but a limited geographical extent, and being continental gifts cannot be older than their source. The alluviums existing ages before a Tertiary deposit may yet be in existence, and may continue in coming ages to throw their treasures into future Tertiaries.

The superposition of the different Tertiary formations are occasioned by the terrestrial movement assumed by this poem. But a few of the Eocene shells in the Paris basin are now found in neighboring seas. The species are pronounced extinct, as it has never been conjectured that the point of the earth's surface occupied by the site of Paris was once at the equator. The Miocene has a less number of extinct species, and the Pliocenes less still. These being sub-continental deposits of a moving continent, the entrance into new latitudes gives new fauna. These Tertiary formations at Paris are named from the comparative resemblance of the Testaceous fauna of each period to that of adjoining seas; and as each succeeding age would bring its formation nearer to the latitude of France, of course the species resembling those of that latitude would gradually increase. This accords with the apparent fact. Mr. Lyell, in his Principles of Geology, remarks :—"The conchological fauna of the Eocene period (speaking of the Parisian basin) indicates a tropical climate; that of the Miocene strata, a climate bordering on the tropics; and that of the Old and New Pliocene deposits, a climate much more closely resembling, if not the same, as that of seas of corresponding latitudes."

But the alluvium, or detritum, is limited to no age; it never ceases to exist upon the earth. We will suppose the earth always divided into two hemispheres—Eastern and Western—both having continents. In process of ages the Eastern will come into the place of the Western, by the route of the North pole. The alluvial fossils which started at Australia would have an overland journey to Ohio and Kentucky, with possible submersions. As this trip would occupy ages, of course, many Tertiary formations would grow up in different corners of those hemispheres, the gifts of their detrital surfaces; but reliques of the oldest, would still remain, and never entirely disappear until buried under oceanic deposits. The alluvial surfaces from which the European and British Tertiaries were derived are now under oceans.

The portion of the Asiatic continent which should pass

under the polar glaciers would carry the marks of its travel on its bosom. The elevated portion would be ground by the ice, polished and striated; the portion not elevated would pass the icy realms beneath the Arctic Ocean, untouched by the ice; and large portion of the Eastern continent might find its way outside of the glacial regions. No fossils could reach us from Asia by the elevated route; but the other routes would give free passage to the Asiatic fossils so abundant in our land. But the drift would involve all. In some portions of America but a single bone of the elephant can be found; in other places whole skeletons are exhumed from ancient Asiatic mire holes and saline bogs, but overwhelmed with drift. The parents of our species should not be looked for in oceanic deposits, but in such ancient alluviums, as are not, with the Tertiaries of past ages, decayed, or buried under distant oceans.

The line of drift is a guide in regard to the earth's motion. The drift stream is wholly in the west. In the east there is no drift, except from mountain centres. The arctic boulders strew portions of Eastern America and Western Europe, but are not found east, or far east of the Ural Mountains.

Gravitation has been regarded as the cause of planetary orbital motions, and the earth allowed no movement not apparently due to the sun's attraction. This fossilized notion has for a long time closed the eyes of philosophers to indubitable evidences of terrestrial changes of position. The simple fact is, gravitation is a mere disturber of motion; that is, planetary motion. The curves about the sun, made by the circuits of the planets, may, as claimed, be in a measure caused by gravitation; but the motions of those planets are due to the force communicated by God. That this projectile force must be recognized, all admit; and it is a force that never is exhausted nor diminished, but forever encounters obstacles and opposition, and still causes the planets to perform their circuits in time, precise and exact. Gravitation may influence the earth in some of her motions; but the great annual movement about the sun, and the daily revolution on her axis, as well as the slow revolution claimed in this article, are motions imparted by the source of power. The motions of the planets, called projectile, indicate a living, active force. If only communicated motions, like that of a cannon ball, continual gravitation would in time entirely overcome them; they are evidently motions imparted by Divine Wisdom. More or less of these motions the prying curiosity of man may discover; but not their source. Divine Wisdom asks no permission of the suns of His universe to move His planets as He pleases, and only sees to it that the opposition of the solar centres works no harm to His system.

Recapitulation of Evidences in Favor of the Earth's Axial Changes.

First—Drift action all on this side of the globe. None in Siberia, as I am told, nor in Asia, except what may be traced to ancient mountain centres.

Second—The drift soil in America and Great Britain includes the remains of vegetation and animals.

Third—All the striæ on the polished rocks of the western continents uniformly point to the arctic regions, varying no more than a slightly spiral movement would allow.

Fourth—Throughout the field of glacial action the ledges are polished and striated, and generally covered with a heavy barden of drift soil. This fact is conclusive, as such result could never have occurred if the glacial action had been in consequence of a cold period that mantled the globe with ice to the equator! The dissolution of such an ice mantle (which, all must see, could never have existed) could not have formed water streams that would have polished our ledges, rounded all the pebbles, and carried such burdens of soil and boulders upon mountains. The drift soil was taken when our continents left the arctic ice, and were moving south, under northern oceans, as land now under the Northern Ocean is taking its burden of drift.

Fifth—Siberia was once a tropical fauna. Its present climate could not support the animals whose remains are found in her bosom. It must have migrated from the south. We find its animal remains now entering the icy realms.

Sixth—We find on our side of the icy circle Siberian remains distributed in the soil; in some instances in scattered fragments; in others, undisturbed. How some would be destroyed and others safely conveyed in submerged routes has been explained.

Seventh—The rock under the arctic regions is fossiliferous; it has its coal beds inclosing animals and plants of a tropical climate. There is no way to account for this, except by the theory here advanced.

Eighth—The great mass of drift soil on our side of the globe is, as a rule, destitute of organic remains. In Asia there is no such a state of things. That is, in Asia the soil is generally normal; in Europe and America, abnormal.

Ninth—That large portions of Asia may show the effects of glacial action is no doubt true; but those effects should be clearly distinguished from the evidences of such action in Europe and America. Asia, after passing from the Western hemisphere by way of the Antarctic, lingered for ages in the region of the tropics, where heat and rains, and the submergences endured, would to great extent obliterate the traces of glacial action. The fact that the characteristics of our ice age do not exist in Asia limits that cold age to our hemisphere, while the other was normal. If the evidences of glacial action and drift were the same all around the globe, there would be no possible standing ground for the theory here broached; and if there are no such evidences, the theory must have firm ground to stand on. And Murchison tells us that there is no boulder drift in Siberia, and that evidences of glacial action are very faint and difficult to be found in Asia.

Tenth—The direction of the glacial movement, uniformly from the northwest to southeast, expending its principal force upon the northwest sides of mountains, even plowing over the summit of the White Mountains, is pretty clear proof that the phenomenon was owing to the revolution of the earth described in the poem.

Eleventh—The boulders themselves, as we survey them from latitude to latitude, declare the truth of the matter. The southern boulders bear the evidences of age, and are nearer the point of disintegration than those of the north. A boulder at New York City would be at least upwards of thirty-three thousand years older than one at Montreal, and the appearance warrants this difference in their ages. No boulder nor striæ mark could live to pass the tropics, unless protected by coverings of earth or water, or unless upon elevations. Of course, it would not be impossible for boulders and striæ marks to pass the equator under certain circumstances; but as a general rule the boulders and marks of ice action in South America and northern Asia must be of a local character.

Twelfth—At the equator are the evidences of drift in the decomposed rock. This fact must be decisive of the question of the earth's movement, as no one will contend that the ice mantle of the glacial age extended over the equator! If there be decomposed drift, in place, in the equatorial plains and valleys, and boulders on elevations, it will not be rash or audacious to ask the greatest of our philosophers to admit that the theory of the poem is right! See authority of Prof. Agassiz, in Silliman's Journal for Nov., 1865, as reported by his son, on the Drift in Brazil. He had seen unmistakable traces of drift in the province of

Rio and in Minas Geraes; but there was everywhere connected with the drift itself such an amount of decomposed rocks of various kinds that it would have been difficult to satisfy anyone not familiar with drift that there is here an equivalent to the northern drift. And Prof. Agassiz, in his lectures before the Lowell Institute, in Boston, in October and November, 1866, on the Geology of the Valley of the Amazon (right on the equator), speaking of the decomposed condition of the rocks in the regions of Ceara and Rio, says he was satisfied that large masses of what we call drift rest on the solid tropical rocks as well as on the rocks in the northern regions; and that these are erratic is plain from the fact that they are not of the same mineral character as the rocks underneath them. And on the slopes of mountains in the vicinity of Mangouva he was struck with the appearance of boulders, first supposed to be local and posterior to the ice age; but on investigation he found these perch rocks, on summits of mountains, of an entirely different character from the rocks on which they rest, and says that they must have been brought by no agency but ice. And the learned Prof. declares that there was a time when not only the northern and southern hemispheres and the temperate zones were covered with fields of ice, but the phenomena extended over the tropical regions !

DARWIN.

Life! it is the gift of God, Creator
Of the world; the great architect who planned
And made the universe.   Design is seen
In every form of life, that plainly shows
The workman.   There was, of organic forms
The first, which, without creative power,
Could not have sprung from inorganic dust.
Evolution! 'tis this that Huxley gives
In place of a creator, with ages
As innumerable as sea-shore sands.
Original germs they were, some two or three,
From which the varied forms of living things
Have developed out, in progressive steps,
By natural laws.   Thus are we made to see
That matter, in its corpuscules, contains
Almost creative powers.   But still, these germs
Of the lowest forms of organic life,
With marvellous tendency to produce
Superior species, and endowed with power
Of propagation, infinite,—whence came they?
Philosopher! can you tell?

PHILOSOPHER.

Life and God, in existence coeval
Are, and both eternal. (¹)

(1) We need not fear to assert that life is eternal That Divine Wisdom ever existed, when there was no life in the world, cannot be supposed. If the existence of life ever commenced, however many ages ago, the fact of such commencement would show that God had existed for an eternity alone in a dead universe ! Life is eternal and infinite. That is, there is no limit in time or space to its existence. A spirit may mount a sunbeam, and in a straight line traverse space eternally, and he will constantly be in a starry canopy like that which encompasses us, and everywhere be the spectator of the abodes of busy life. But we have no knowledge of life outside of organization, or organic forms. These forms are almost endless in their varieties, and each variety is a marvel in mechanism, and is adapted to its locality with perfection. One law in regard to the prevalence of organic life seems to be universal. Wherever there is food for living organisms, life will abound ; and the forms are adapted to the food destined to its support. The worm in the mud, the serpents in the waters, the innumerable insects, and insect devourers, and every other living thing, are shaped, organized, and dispositioned each to the sphere of its nourishment.

Man may not see the wisdom in the authorized existence of beings that seem to him pernicious, and deadly hostile to him ; and is mortified to see evidence that possibly all things were not designed for his use. But, in this universal prevalence of life, the wisdom is infinite. Life is joy, and every jot of organic matter in the universe is subjected to its emotions. This was always so ; this always will be so. And he is a brute, a rebel against the Divine Infinite, who will unnecessarily destroy or molest, any living thing.

CHRISTIAN.

Did life here exist
When the earth was but a fiery vapor?
Or an incandescent mass of matter?

PHILOSOPHER.

That earth was ever in a gaseous state,
Is mere conjecture; and philosophy
With conjectures deals not.(2)   We think we know

(2.) It is very generally believed that our solar system had a
beginning. This is probably a popular error which will never be
eradicated. The human mind is wisely so constituted as to make
it distrustful of new opinions, and novel facts. And, furthermore,
as Divine Wisdom, by renewals, keeps the human race ever young,
it is but natural, if not important, that human beings should regard
everything in the world as in its youth. Nature certainly diverts
our minds from the truth in these matters; and as man's best good
consists in a limited vision, and limited range of thought, I do not
expect many will accept the views of the poem as to the age of the
world and the things therein.

The nebular hypothesis is quite commonly received by men of
science.   The probabilities in its favor, as worked out by La
Place, are strong and convincing, on the hypothesis that the
system *had any beginning at all.*   If it be allowed that the solar
system is eternal, there is not the least significance in all or any
of the particulars so much urged by astronomers in favor of the
nebular theory, nor can those particulars raise the slightest inference
that the system had a beginning.   That all the planets and their
respective satellites move and revolve in one direction, may indicate

That matter is eternal.   This premised,
We see not why the universe of worlds,
As they now in systems revolve in space,
Should not be eternal, too.   And if so,
Why of the solar system make exception?
That these spheres from old to new bodies change
We have no knowledge ; nor have we knowledge
Of any law for such a transformation.

a community of origin, *if origin there was;* but these particulars do
not indicate an origin.   The converse is true.   That a group of
bodies is eternal is no sort of reason why it should not be systematic.

There are many philosophers who concede the existence of
Divine Wisdom.   Matter is usually thought to be eternal, and, of
course, it must be admitted that the material of the universe never
had a beginning.   So far wise men agree.   The presumption is that
the solar system never had a beginning, but many believe otherwise
without any evidence.

That the matter of the world remained in a chaotic state from
eternity down to within a comparatively recent period, when it
worked itself into suns and planets, is but mere conjecture.
However regular, however systematic, however concordant and
harmonious, our planetary families may now be, is not the slightest
reason for thinking they must have had a commencement.   The
regularity and harmony noticed most truly indicates wisdom, but
we must recollect that wisdom had no beginning.   The material
universe could not exist except in accordance with the laws of
Divine Wisdom, and it is childish to suppose that these laws are
of recent origin, and that all the works of nature have but
recently commenced.

Much is said about nebula.   If it exists it may not be matter in

CHRISTIAN.

Surely this earth was incandescent once,
And destitute of life; or the science
Of geology falsely teaches.

state of transition to worlds. To support the nebular theory old worlds must be resolved into nebulæ, and thence into new worlds. This is certainly utterly impossible according to all the laws of matter with which we are acquainted. If we do not know certainly that such a transformation takes place we should pronounce it impossible.

There are things connected with the solar system not accounted for by, and are in conflict with, the nebular theory. The nebula, in space, may have had a revolution about its centre, from which the motions of the planets about the sun were derived, but where did the planets and their satellites get their revolutionary motions on their respective axes? Something more than evolution would be required to account for all the motions and habits of the planetary orbs. But the comets arrest attention. Were they segregated from the nebular mass that produced the sun and planets, and do they belong to the solar system? They regularly circulate about the sun, although the sun is rapidly moving through space, and they return faithfully to our solar centre after an absence of many hundreds of years, and after visiting other solar systems. The truth is the heavenly mechanism is too complicated for the supposed nebular origin, and the nebular theory is but a cloud of philosophical nonsense. This theory in the threshold is met by impossibilities, and is a satire on human reason.*

* See able editorial in N. Y. *Tribune*, Oct. 9, 1877, entitled, "A Great Theory in Danger."

PHILOSOPHER.

The science of geology is young ;
The earth is old.   And those azoic rocks
Which have been thought merely foundation stones
Of life's temple here on earth, were, no doubt,
Themselves the abodes of life, when in their prime.
And the direful reign of carbon, also,
Is as imperial now, as in the days
Of those old hard-shells, orthoceras called.

CHRISTIAN.

Why are tilted the strata of the earth,
If there's been no contraction of its size ?
And how contracted, unless its temperature
Has been reduced ?

PHILOSOPHER.

The present must explain the mystic past.
Strata now, in every sea and ocean,
Are being formed ; and there's but little doubt
But time will see them into mountains,
Hills and valleys, by some convulsion, thrown.([3])

(3.) I had supposed that the elevation of continents above, and
sinking below, oceanic levels, by a natural, gradual process, had
long since become recognized phenomena. But I see that, at our

CHRISTIAN.

I doubt not, causes for such upheavals
May have existence.   But when we behold

new Academy of Sciences, last April, Prof. Alexander Agassiz re-
jected this theory.   He said :—"There is an entire want of evidence
that great continents existed where oceans now are."   In maintain-
ing this, the son follows in the footsteps of his distinguished father.
It would appear that this position of the great Agassiz shows how
a person may be a very great naturalist, and yet not distinguished in
other departments of science.   He may enrich the world with valuable facts, and still not be the soundest of reasoners outside of
his specialty.   It is almost certain that Agassiz, in pondering upon
geological subjects, could not make allowance for the immensity of
time necessary for geological changes.   He sees, as he says, no
change of fauna on the ocean beds for thousands of years, and hence
infers that there are no upheavals.   The geologist takes in more
extended periods, of which he has ample evidence.   All agree that
at some past period all our present continents were under the oceans,
as the immense deposits on them must have been acquired when
submerged.   These deposits are many miles deep.   If there were no
continents in the world, when our present continents were submerged, taking their immense burdens, where did these burdens
come from ?   Indeed, there may have been "an entire want of evidence" in the bottom of the ocean where our distinguished friend
at Cambridge was looking; but, if he ventures opinions in geology,
he must look further.   It is, besides, a recognized fact that none of
our continents maintains a steady level.   They are all changing
slowly, but continually; and that very ocean bottom explored by
Prof. Agassiz, if properly watched, would, no doubt, be found
gradually approaching, or receding from, the surface.

The earth's tilted strata, from east to west,
And from north to south, with a general strike,
We must seek, for results so uniform,
Some single cause.   The future, it is true,
May upheavals bring ; but will the strata,
Forced up by local causes, in their trend
Be as uniform as those that now are here ?

PHILOSOPHER.

Contraction and expansion, are causes   •
For upheavals ; but 'tis the solar heat,
With gaseous forces in the rocky depths,
And not abatement of internal fires,
That cracks up earth's stony vest, or shell,
Causing fractures, at right angles running
With the equator.   I know, it is claimed
That earth was once a fiery gas, or mist,
Changed to its present state by loss of heat.
This was an old school suggestion, and made
Before the chemist came.   The modern sees
That the solar system, if all in gas,
Spread out in space, would not, in temperature,
Range very high.   No law is known to man
By which this earth could in space exist
In gaseous state, though great astronomers
And geologists of note, have thought so.   ´

CHRISTIAN.

At the equator, where the solar heat
Descends in force, the sun's power might be great;
But upheavals, in every latitude,
From pole to pole, are seen.

PHILOSOPHER.

Very true; but the equatorial line
Is forever changing; and that, by steps
So slow, as to escape the observation
Of the astronomer.(⁴) The polar ice

(4.) Mr. JAMES CROLL (*Climate and Time*, London, 1875), brings
forward evidences of former glacial periods, some ten in all, with
high authorities, which he collects in his book. He shows that the
Cambrian period encountered glacial action; also the Silurian period;
also the Old Red Sandstone; also the Carboniferous; also the Per-
mian; also the Oolite; also the Cretaceous; also the Eocene; also the
Miocene; and we since have had glacial action in the recent, or Post
Pliocene period. Some English gentlemen now claim they have
evidence of man's existence prior to the glacial period, that is remains are found under the boulder clay. Mr Sidney B. Skertchly,
of Brandon, is decided in his opinion that he has evidences to this
effect; and Prof. James Geikie and others are of the same opinion,
although there are doubts expressed by Prof. Dawkins, Mr. Evans,
and Prof. Hayes. However, Mr. Skertchly's evidences, resulting
from recent explorations, as we see by a notice in *Nature*, are not
before the public. But these gentlemen, or some of them, claim an
inter-glacial period of warm climate, and that these remains belong

Alone is stationary ; under it
The earth is moving with regular pace,
In perhaps a spiral course, which, in time,
All its parts must bring to the icy zone.
There, at the poles, are the mills of the gods
Which grind so slow and fine.(⁵) Our continents

to the inter-glacial period. Mr. Croll, in his *Climate and Time*,
brings forward much evidence of an inter-glacial period, by the
remnants of the Cromer Forest, peat beds, roots, and other reliques
of vegetation, and remnants of animals. There is something unaccountable in the pretense of two glacial periods. But boulder clay
is the product of glacial action, no doubt; and if, between two beds
of it, remains of trees or animals are found, what can be said against
the proposition of two drift periods ? So far as man's work is
claimed to have been found, Prof. Boyd Dawkins will not admit
further than proved; but the occurrences of animal and vegetable
remains within the genuine drift, he will not deny. And how is he
going to manage two glacial periods ? After struggling with the
problem for some time, I think he will concede the claim of this
poem, that these reliques of vegetation and animals came along
under the polar glaciers, from Asia, pretty scant and much used up,
no doubt. I trust he will think the matter over carefully, and be
really liberal and free from bias in the premises. It is a great burden
to carry two glacial periods so near each other; and it is not easy to
fancy the earth going into the elliptics figured out by Mr. Croll.
There is something hysterical in the imagined performance.

(5.) As our formations work downward, and finally are changed
into granite, and hove up in mountains, some agency becomes necessary to unlock the vegetable treasures they contain. In the mica

In those mills are made, together with our soil.
Broken rock, dirt, dust, and rounded pebbles
Are brought forth, with the emerging continent,
And lugged away by ice-bergs, to be strewn
Where ocean currents spread them o'er the earth.
On our hills, in our vales, we see the marks
Of polar ice; the ledges rounded, smoothed
And polished by the comminuting mass.
The striæ left upon these rocky points,
Point to the polar realms.([6])

and felspar is from eight to ten per cent. of potash, and other rich
elements are included in the granite bosom. At each revolution of
the earth under the polar glaciers there is great comminution of
rocky material, and vast quantities of elements that promote vegetation set loose. Sunshine and rain at the equatorial point, and grinding ice at the poles, with great rapidity break up and disintegrate the
granite mountains that otherwise might indefinitely hold locked up
the vegetable treasures of the earth.

(6.) Since the first edition of this book was printed, George II.
Darwin, F.R.S., an accomplished son of the great naturalist, has
presented to the Royal Society of London a paper showing that the
evidences of glacial action on our continents are owing to a movement of the arctic ice, occasioned by a change of the earth's axis.
That there was no ice age in fact. That this change of axis is
owing to an oscillation, the cause of which he explains, and which
oscillation may carry the axis away some fifteen degrees. The
inclination to the ecliptic is not changed. Of course the theory of
this brilliant young man encounters geological objections. Of course

CHRISTIAN.

The polar zone
Is then receding from the western world ?
But if the Atlantic basin has moved
From beneath the frigid zone, why are found
Those fields of coal in the far distant north ?
Those carboniferous beds, as we are told,
Numerous plants contain of tropic growth.
Then why toward the equator should they move?
And why those fossil forests in the north,
In the realms of perpetual frost and ice,
Unless the earth, if it does move at all,
Is creeping the other way ?

PHILOSOPHER.

First tell me
How came those tropic plants, thus fossilized,
In the frigid north ? Around Baffin's Bay
The gigantic fern of an age remote,

there are objections. But science will yet demonstrate that he is
right in his main proposition. The glacial period exists now as
much as ever; and the change of the earth's axis is the true cause of
glacial marks on our continents. There is not science enough in
England or the world to overthrow these facts, for facts they are.
The motion which I give the earth is more revolutionary than that
proposed by Mr. Darwin; but I live in a more revolutionary country
than his, and can afford to do so.

Carbonized in its native soil, is seen,
And speaks like history of a former age.
Tropic plants must within the tropics grow.([7])
Some have said that, warmed by internal fires,
The earth was once a sort of hot-house globe,
At whose poles no chill atmospheric cold
Could check the growth of vegetation.
Then was the air warm, indeed, for, aloft,
In polar regions, did aspiring plants

(7.) Sir John Lubbock seems to be the only geologist who sees the
impossibility of a coincidence in time of the principal great geological formations, as found in widely separated parts of the globe; and
he boldly and conscientiously announced his convictions, although
they subjected him to criticisms. Formations are determined by
their faunas and floras, and these must always be identical, no matter in what part of the globe the formations are deposited, according
to the teachings of geology. The great Mr. Lyell struggled with
seeming anguish to get over the difficulties involved in the position,
and the faithful and bold Sir John Lubbock confronts it. It is
pleasant to encounter and combat a popular error; but the first
sensations, on seeing how tenaciously the popular mind clings to its
errors, are painful. More study and reflection, however, reconciles
the champion of truth to his condition, as this persistence in fixed
ideas may be in consequence of a quality of the human mind that
is productive of great good. If men could easily be changed in their
opinions, there would be nothing stable in the social fabric. Social
fabrics are sometimes founded on error, and are prosperous, peace-
able and happy, till some agitator puts them in commotion by the
promulgation of truth. Promulgating the truth is sometimes a crime,
and I hope to be mercifully treated.

Spread to the breeze their leafy branches.
A summer sky was there, to furnish clouds
For genial showers. To impart a warmth,
To such extent, to the polar breezes,
How hot should be the earth beneath? Not less,
At least, than broiling point for fish or steaks.
Evaporation must ensue, and drouth.
As at the equator, roots could not dive
For moisture, for they would encounter steam.

Fossils of tropic plants and animals
In many latitudes are found, which shows
That terrestrial changes have oft occurred.

CHRISTIAN.

We have never heard that the latitude
Of any place on the earth, has been changed
From the earliest ages.

PHILOSOPHER.

  Earliest ages?
How long is that? History backward reaches
Three thousand years, almost; whereas, a tree,
I have heard it said, may four thousand live.
Such terrestrial changes should be looked for
In the record of cycles, only; not

In human history, which is limited
To events of years. When the arctic land
Was within the tropics, as once it was,
Its departure was very slow, of course,
As on its bosom, the accumulations
Of a vast geologic age, it bore.
Which way does the terrestrial movement tend?
The question may well be asked; for no change
Since the dawn of modern science, will point
Its direction. On the Atlantic's shores,
Both east and west; in Eastern America
And Western Europe, the earth's rocky ribs
Show the grinding power of icy mountains.
And then, upon the globe's other side,
Throughout Siberia's arctic realms, 'tis said
The frozen elephants with force proclaim
A warmer climate at some prior age.
Such facts may show that our Atlantic land
From Asia came. The Atlantic basin
Was a central point of the polar land
That passed beneath the arctic snow and ice,
With water level changing all the time.
The carboniferous treasures which are found
In the northern world, must Asiatic be,
As also the fossil trees. But the trees,
Upright, as now they stand, to have passed

Beneath the polar ice, in the ocean
Must have been submerged; or they may have grown
When the gulf stream bore further west; but plants
Of tropic growth, must from the tropics come.

CHRISTIAN.

If this be so, there is no primitive
Formation !

PHILOSOPHER.

They view not the rocks aright
Who to the old Silurian system point
As holding the fossilized first parents
Of earth's countless species; for that system
Was in the deep bosom of the ocean
Formed, and was not a fauna rich in life.
But, while that formation was in progress,
A tertiary, and alluvium, also,
Must have been extant, freighted with the life
Of the present day, which are now, no doubt,
In the ocean's deep bosom buried.
No formation o'er all the earth extends;
And as each grows up in the ocean's depth,
Another must disappear. The faunas
Of all formations must depend, of course,
On their position in their watery bed.

And the Cambrian, 'neath fifty thousand feet
Of brooding ocean, would not be the home
Of the prolific broods of shallow seas.

CHRISTIAN.

I must beg
To call you back to your definition
Of philosophy; to what it deals in.

PHILOSOPHER.

It deals in facts, and facts alone.

CHRISTIAN.

Then hold !
Should we not stick to facts—to what we know?

PHILOSOPHER.

There are certain known invariable laws
Of matter, a knowledge of which makes up
The sum of our philosophy. How moves
The earth to warm with sunshine all her sides,
Or how, by wasting fires beneath—by rains
Or corroding frosts above, the strata,
By disintegration are worn away,
Is speculation foreign to our theme.
The tracks of time may not be rightly nosed

By the keenest scenting philosopher
That ever embarked in chase of nature.
But this we know, by fair deduction
From well-known laws, that, as time began not,
And as the world is as old as time,
All things brought forth by time's effluxion,
That in the future, we see must occur,
Must be recurrences of like events
Repeated *ad infinitum* in the past.
That every strata now upon the earth
Must some day be into new strata formed,
Suggests the changes of the past. But then,
We must bear in mind the controlling fact
That organic structures are coeval
With the universe, and God himself.   ·

CHRISTIAN.

Organic structures embrace all species
Of plants and animals. If eternal,
Why are some extinct?

PHILOSOPHER.

Forms alone may change.(8)
What ends, began. If from primordial forms

(8.) The world is looking anxiously to evolution as a possible solution
of the problem as to the origin of animals on the earth. The doctrine

Species have come by evolution, then Primordials only must eternal be.

is not proved. It has been canvassed and attended to with remarkable candor by all classes of people, of all religious denominations. The sincere Christian concedes that God, in the creation of terrestrial beings, worked through natural laws, and is perfectly willing to accept natural necessity as his instrumentality, if proved. Bringing a germ from the dust by action of the principle of natural necessity would be a method of the operation of the Divine will, and might, with great propriety, be called creation. Mr. Huxley, in his lecture in New York City, in 1876, on the horse, claimed to prove the doctrine of evolution. He pronounced it as well established as the Copernican System, and based his evidence on the changes said to have taken place in the horse. Although Mr. Huxley is one of our distinguished naturalists, to whom science is indebted for valuable labors, it seems to me that he made a mistake in his reasoning upon this subject. If there is anything to evolution, it consists in the origination of higher species from lower, and lower from protoplastic matter. By this means new organs may be brought into existence, and one species transmuted into another. But anything less than this cannot be evolution; that is, changes induced by circumstances, not going to the extent of changing the species, can hardly be evolution. The absorption or decay of organs no longer used is not such. No one will claim the fish taken from artesian wells, whose eyes are obliterated from long disuse, is a case of evolution. The principle of reversion would in time restore the eyes under proper circumstances. It seems that Mr. Huxley has found that the horse of early ages had not only one main hoof, but also four minor ones, and also had two bones in each fore and hind leg; while our present horse has but one hoof on each foot, and the *u'na* and *fibula* are mostly obliterated. It is a case of the loss of parts by

We trust there's none extinct. Every species,
In its structure, and faunal adaptation,

disuse, but not to an extent to change the species. But the learned
professor remarks:—"Hence it follows a differentiated animal like
the horse must have proceeded by way of evolution, or gradual
modification, *from a form possessing all the characteristics we find in
mammals in general.*" I supposed he was a horse, as found in the
Eocene. If the orohippus of Huxley was not the same as our horse,
except in particulars named, we are misinformed. The principle of
reversion makes all organs lost by disuse the animal's real possession; they are merely laid aside till wanted, and circumstances may
yet bring pomp back upon all his toes.

Mr. Huxley, I think, did not give the size of his Eocene horse.
Prof. Marsh says it was of the size of a fox! The Shetland pony,
then, is a connecting link! But the little Shetland horse and our
noble animal are cotemporaries. They are varieties of the Equine
species. That little thing of the Eocene may have been a variety;
but we can hardly be sure he is the progenitor of our horse. Mr.
Huxley himself puts some horse fossils out of the line of descent;
why not a layman make exclusions? For instance, in Europe are
the bones of the *Hipparion;* he says, in his New York lecture,
this "three toed horse in fact presents a foot similar" to Mr.
Marsh's American *Protohippus*, "except that in the European
Hipparion the smaller fingers are farther back, and the lateral toes are
of smaller proportional size." But, since those lectures, Mr. Marsh
has found one more, another horse, the Eohippus, with all the toes
complete. It begins to look as if we have in our Western Tertiary
the fabled Mare's Nest, and may "begin to look for eggs." "There
is," said Mr. Huxley, "this Pliocene form of the horse (*Pliohippus*);
* * * Then comes the form which represents the European
Hipparion, which is the *Protohippus*, having three toes and forearm

Discloses perfect wisdom; a wisdom
Not evolved from, but coeternal with,
The organic being. Species do not die out,
But change, if into other faunas forced;
So that the links of life new forms may take,
But the chain itself is never broken.
If a creator, and he e'er began,
Then to God himself, an eternity   .
Was lost. He began not. Organic forms,
With power of sense and thought, were either made
Or are eternal.(⁹) But suppose we say

and leg and teeth to which I have referred, and which is more valuable than the European *Hipparion* for this reason: it is devoid of some of the peculiarities of that form—peculiarities which tend to show that the European *Hipparion is rather a side branch than one in the direct line of succession.*" If one set of horse bones can be excluded from the line of descent, who can say certainly which shall be included ?

(9.) The earth, when habitable or capable of sustaining any one species. must have been in a condition to receive and support all the species ever known to have been upon it. If animals came by special creation, they may have come all at once; if by evolution, every species may have been evolved at the same period. For argument, let us concede that the earth came out of incandescent matter, and in time attained a habitable condition; and also concede that the species of animals must have had points of origination. Evolutionists claim the transmutation, by natural selection, or influences of circumstances, co-operating with a tendency to variation, of one

That Darwin's branchifers the power of thought
And sense embrace.　By this we little gain.

species into another; and this accounts for the numerous species—a
few low orders having given birth to all.　In such transmutation it
has been suggested that some individuals were changed, leaving
the old species still existing.　The evidence relied on for this doctrine
is the discovery in the anatomy, organs, faculties and characteristics
of the higher species, of a similarity to the organizations of lower
species.　For instance, in man are the rudiments of muscles not used
by him, but in full vigor in lower animals.　Reversion will furnish
man sometimes with organs or traits peculiar to lower species.
These are held as strong evidences that the lower animals are in the
line of man's descent.

If we look to evolution as the source of life, we are told that all
the species came from a few original germs; that these germs were
derived, in some unexplained manner, from inorganic matter.　To
carry forward the evolution system, we must suppose that all life
below the mammalia appeared first.　The highest order of animals
came from the lower, by gradual steps, the connecting links having
had existence, though not now in sight.　If developement be the
true system, of course the *herbivora* preceded the *carnivora*.　The
abundance of wasting herbage gave a call for devourers, and hence,
came into being the giraffes, camels, elephants, moose, buffalo. elk,
etc., not naming domestic animals.　After these herb-eaters got
possession of the earth, without an enemy, their increase was so
rapid that, without the interposition of some check, destruction must
soon overtake them all.　There was a plump, positive necessity for
a means of keeping down the increase of the *herbivora*.　This was
supplied by evolution, as we are told.　As the food began to grow
scarce for the fast increasing mouths, and the gentle grass-eater began
to feel the effects of famine, he must have longed for claws and teeth

Whence those branchifers? If self propagating,
Organic, with sense and thought indued,

with which to devour his kindred; and thus, by natural necessity,
was he changed into a carnivorous creature. Or, maybe, the great
necessity of the occasion, penetrating by its mystic influences all
animate nature, caused a spontaneous production from some lower
form, in whose organism the proper tendency existed, or from protoplastic matter, which had no tendency at all, of the great variety of
carnivora, such as tigers, hyenas, lions, leopards, bears, etc., which
have, indeed, saved the herbivorus herds from extinction.

But it is not claimed by Mr. Darwin, so far as I know, that natural
selection, the great hand-maiden of evolution, can act on inorganic
matter; nor can it be operative by any theory unless the heart, which
is to inaugurate the transformation, feels the necessity of the occasion
It is the pressing need of the hoof that causes its developement; the
felt need of the claws that calls them forth. The bird gets not the
beautiful plumage that promotes sexual selection until its want is
felt. In short, fleet limbs, swift wings, strong muscles, protective
colors and bony shields, are acquired by natural selection, because
their necessity has been felt, and are the slow and gradual acquisitions of ages.

In coming to our relief, as an explanation of the origin of animated
nature, the evolutionist who bases his theory on the gradual changes
effected by surrounding conditions, must satisfy us that species, in
their commencement, were not perfect. We are all taught by the
greatest naturalist the world, since the historical period, has produced, how marvelously individuals may be changed by change of
conditions. But, as to the species, does fossil lore teach us that any
one was ever imperfect ? We surrender at once, for the sake of
argument, to the proposition that the turtle emerged from some
amphibia or inferior animal; but did ever one exist without his

The embryo of superior races;
Although of form most simple in the scale

shell ? We may suppose that the lizard took wings and feathers and became a bird; but was not the first bird, at the time of the transmutation, just as perfect as any of its descendants now upon the earth ? Does paleontology furnish evidence that, at any prior age, cranes, storks, eagles or ostriches were substantially different from those now in existence ? that is, was there an age when the crane had not long legs, the stork long bill the eagle his powerful talons ? Is there evidence that species have changed their distinctive characteristics ? Without this evidence, evolution by gradual changes effected by surrounding conditions, cannot be maintained. The permanency of types is attested by Mr. Huxley himself, as he cites examples, from the oldest formations, of species identical with those now existing; and Mr. Darwin, after showing the modifications of parts, remarks:—"Each organism still retains the general type of structure of the progenitor from which it was aboriginally derived." Then, passing over the variations flesh is heir to, and coming to the species themselves, we see that, if they ever had a beginning, it was from a progenitor, and not from an interminable period of gradual changes. It is true Mr. Huxley picks up the bones of the horse of the Eocene, and finding that his hoofs and legs are in some respects different from those of our present horse, proclaims his problem proved. But his diagrams and lectures only show his *Eohippus* more perfect than our present *Equus*. He was really armed with five hoofs instead of one ! and his legs had two bones instead of one! Disuse, lone, has caused temporary disappearance of useless members! If this is evolution, it evolves backwards ! Evolution may be one of the eternal laws of Divine Wisdom, and probably is; but the evolution that works simply by force of circumstances must be destitute of wisdom or purpose, and we must not expect too much of it.

Of animated being, their origin
Is a mystery still. Whence came their germ?
Could it ever have had a starting point?
Was it, by natural selection, teased
From Huxley's mud or jelly.([10]) Or, in short,

The presumption is, that this earth has been here eternally, and consequently life upon it had no beginning. But if man and the domestic animals ever had an introduction upon the earth, there is no evidence that they have been changed since. The sheep, the horse and the cow, were never different from what they are now, in their main characteristics. Without man the natures of domestic animals would not have been developed, as evidently it is a principle of Divine Wisdom that they shall be dependant on man for the perfection of their natures. In nature, without man's care, they would become extinct, or nearly so; but under it the earth is filled with them. The butcher simply acts like the nursery-man, who cuts away many beautiful, healthy branches, which bleed and perish that the remainder may be more fruitful. The utility of animals to man, and man's interest in their preservation, are points of wisdom which could never have had a beginning. Divine Wisdom only acts by laws. When we seek to trace this wisdom, in certain directions, we are sure to be deluded, as it is evidently the purpose of nature to mislead the mind of man in all attempts to look behind the curtain. No doubt man's happiness is promoted by the delusion. There is an observable natural bent and direction of the human mind in cer. tain matters, and it is fortunate that man is governed thereby.

(10.) Mr. Huxley concedes the possibility of the origin of organic forms from protoplastic matter. Such origination must be independent of natural selection, unless the latter acts upon inert matter. It

Did it come from a creative hand?
'Tis vain to dwarf its origin, and say

comes to this. The earth, or a particular fauna, may be in great need of some peculiar animal, and this need of inanimate matter must be pressed in as the moving cause of the origination of a germ from gellatinous sediment; and this we call evolution by surrounding conditions. Let us suppose an occasion arises for a new species, and such species arises from protoplastic matter, is it right to attribute the operation to circumstances? The necessity of the locality has its own way; and as it gets its germ which, perhaps, carries in its bosom the germs of higher species, why should not that germ be invested with all the faculties, including instincts and tendencies, necessary to protect it in its future career? If circumstances can evoke species from the dust to answer particular spheres of animal existence, why should it bring them forth half finished, leaving them to the experience of ages for the acquisition of the instincts and faculties necessary for their existence and comfort? If germs really do originate, the reasoning which invests them with necessary instincts is sound, although it necessitates a law of Divine Wisdom in the operation.

Our philosophers are giving to environment, or circumstances, the attributes of force, when, in fact, they are but the disturbers of forces. We can see how they may bear on individual action, but sometimes think too much is attributed to them. The instincts of animals are big stumbling blocks in way of evolution by circumstances, as no species could well have lived till indispensable instincts are acquired. These instincts are not credited to wisdom in the principle of animal life, but to an inherited influence acquired by the long experience of ancestors. If we concede that the species had a beginning, evoked from low germs by purely natural influences, the first generations of the poor creatures, before the proper stock of

'Twas developed out from some still lower germ ;
For faculties of thought were never formed,
Nor were ever organic forms brought forth,
From senseless elements, lest this flora
Or that fauna, should be a barren waste
Without appropriate occupants.([11])

experience had conferred its impressions, must have fared badly, as we do not see how life, without such instincts, could possibly have survived. That certain tendencies to action acquired by animals may be transmitted to their offspring is a well-known fact; but this is because the acquired habits are in the line of the natural and peculiar faculties of the animal, and are transmitted only in consequence of their being the developement of the inherent, original instincts or sphere of organic action. And in this line only can there be variations by natural tendency, or that natural selection would not exterminate at once.

(11.) If species are produced by evolution, natural selection, though a favoring, and to some extent a developing, influence, cannot be the moving cause. It must have behind, or in it, a propelling force, displayed in the main characteristics of life. The tendency to perpetuate itself is one of those characteristics. The means and tendency in this direction are strikingly noticeable for their universal predominence, and certainly such traits could not have been acquisitions of experience, but are as old as life itself. Perpetuation seems to be the single end and aim of all organic beings, as though each lived for this alone. Darwin's writings are full of evidences to the point. His little, young coccus, he tells us, takes but one meal. By her probosis she attaches herself to a plant, sucks the sap and never moves again ! A short life, indeed, but long enough to pro-

CHRISTIAN.

And still
It is thought that life is but in its dawn
Upon the earth.

duce her brood. The life of the silkworm is brief, spent in preparing a deposit for her eggs, when she dies. Many insects live barely time enough to fix deposits for eggs, inclosing them with food for their posthumous young. As these beings never see their young, how could the lessons of experience ever have impelled them to the performance of such duties ? Unless wisdom shapes and governs organizations, how can we account for the operations of what we call instinct ? The little insect that incloses its eggs with dead flies, in a mud cell which it carefully constructs in some safe place, has provided amply for its progeny. But more complicated beings, of longer lives, and, perhaps, of a higher order, are obliged to make proportionately longer provision for their offspring. But the lives of the highest orders of beings extend not beyond the period of activity of their faculties of reproduction. As soon as the offspring, which require years of care and protection, attain points of safety, the parents retire from existence. As reproduction by generation seems the great line of activity of all organic beings, no individual is allowed to survive the period assigned him for such purpose. Natural necessity, unless divinely wise and thoughtful, would not have put such limitations on the powers of creatures that she might coerce out of the dead elements. It is difficult to see how the power of reproduction could possibly be conferred by simple circumstances. That natural selection is one of nature's powerful instrumentalities, Mr. Darwin has abundantly shown; but he has coupled with it a natural tendency which really performs the duties of Divine Wisdom, and, of course, co-exists with it. Geologists insist on the occa-

PHILOSOPHER.

Yes; and 'tis because there is and can be
No record of the past. Those rocky leaves
Of earth's history, which make some note of time,
Are scanned with wonder by one whom the thought

sional loss of species, and if this be so, there must in nature be a
compensating force that replaces them. If species exist by laws of
Divine Wisdom, which of course are eternal, and the full, ample and
wonderful provisions for their preservation and multiplication are
due to those laws, it would be very strange, indeed, if laws were not
in existence to compensate for their accidental or incidental extinction. If species are liable to become extinct, the incident was not
unforeseen nor unprovided for by the eternal laws of being. The
duty of supplying them would not have been thrown upon chance;
and we may accept the tendency to variation, with natural selection
as the chosen instrumentality, if it can be proved; at all events the
compensating power reposes in life itself, which is armed with
ample resources for all natural emergencies.

Men are curious, and often inquire into the purposes of organic
life. It is probably enough for us to say that the happiness and joy
produced indicates Divine Goodness as well as Divine Wisdom.
That life's joyous spirit may be exerted to the fullest extent, the forms
of organic beings must be vastly varied, so that numberless species
may inhabit the same terrestrial domain. Each organic form has its
peculiar nature, in the indulgence and gratification of which its highest joy is attained. Natural selection, as set forth by Mr. Darwin,
it seems to me could have no tendency to draw a species from its
nature, carrying a being out of the line of its activity and natural
career, but must be limited to the derivation of beings with capaci-
ties, in the spheres of the originals, so enlarged as to constitute a new

Of eternity confounds. Life, we see,
Is combustion, an all-destroying flame,
That leaps from form to form, and leaves behind
But an ashy mound. Eternal ages

species. But the evolutionist will say that the lizard and the bird
may inhabit the same climate, and subsist on the same food, only
that the bird has superior means of helping himself to his meals, and
can, therefore, be multiplied to a greater extent. This would be a
good reason for making birds out of lizards, provided that the lizards
were the most ancient. How circumstances could ever work such a
transformation, unaided by a working law of the creature's being, is
quite difficult even to imagine.

But Mr. Darwin does not credit natural selection with the origin
of life. He finds in nature a tendency to variation in species, and
these variations, he holds, are perpetuated by natural selection.
Natural selection is, then, merely the preserver—not creator—of
varieties or species. All the peculiarities constituting a variety must
be inherent or original. Surrounding conditions may destroy, but
cannot create, a species. If a change of one species into another,
according to Mr. Darwin's system (and it seems a sound one), it
must be accomplished by a natural tendency to the change—the
change, in fact, bringing the species into more favorable circumstances. The tendency to variation is a marked characteristic of
species. But does this tendency go beyond producing varieties ?
It was the opinion of Cuvier that it does not. Of course there is
nowhere in nature witnessed a tendency in a species to vary, except
in the line of its organization. Any other change would result in
destruction, by the operation of natural selection. A forced change
by adverse circumstances would, by a diminution of strength and
vigour, take away protection against enemies. The strong alone
survive. But the change being simply a variation attended with an

Have seen this earth replete with teeming life,
And still the flush of youth is on her brow.
He who was, and is, and ever shall be,
Is ever young, and in the present lives.
Youth is nature's aspect—eternal youth.
There's nothing old; for with careful fingers
Death, creeping on life's footsteps, erases
Every trace of past existence. 'Tis true
A few fossils, hid away in ocean caves,
Are by some chance convulsion thrown to sight,
To keep a trace of life in mortal view

increase of force in the organic form, the vital power is increased. Each species is adapted to its surroundings, and its greatest vital energy is realized when the adaptation is perfect. When any of the necessary circumstances are wanting, the power to maintain life amidst enemies is diminished, and extermination the result. Natural selection would be a dire enemy of every change of organic forms that did not spring from the immutable laws of life. And Darwin, who seems to see and think of everything connected with life, while tracing out the changes induced by natural selection, puts in the proviso:—"We should bear in mind that modifications in structure or constitution, which do not serve to adapt an organism to its habits of life, to the food which it consumes, or passively to the surrounding conditions, cannot have been acquired by natural selection." And is it not clear that this limitation of the work of natural selection, prescribed by the father of the system, precludes the idea that the transmutation of species—that is, the transformation of a grass-eater to a flesh-eater—could ever have been caused by surrounding conditions?

For a few odd millions of years, to shock
The spectator, and fill his soul with awe.
But in her living walks, by a system
That never changes, nature permits no show
Of decrepitude or age to mark her reign.
Every mortal being has a time for life,
And each nation, tongue and people, a day
And generation. Three score years and ten
Are counted the span of a human's life;
Two thousand years a people's age. No tongue
Now numbered among the dead, ever served
For life's sweet uses for a longer term;
Nor will one ever for a longer serve.
Had some wild poet, when Antoninus
Reigned o'er Rome's imperial realms;--when, indeed,
The whole civilized world itself was Rome,
Her laws, letters and language extending
From Asia's borders to the British Isles;—
To that imperial lord have said, "My Lord,
In five hundred years the Latin will cease
To be a spoken language;" his lordship
Would have shown incredulity, no doubt.
No doubt but peoples spring, as by a sort
Of evolution, from human masses
Derived from disintegrated nations;
But human history goes not back so far.

In ten thousand years from now, no knowledge
Of any art, tongue, people or nation,
At present existing upon this earth,
Will survive. Aye; let us look to the time
When from the Atlantic a continent
Shall appear. History will repeat itself.
Prophetic races first come forth, of course;
Then, by slow degrees, midst bloody conflicts,
Civilization wins her transient reign.
The number of dead and forgotten tongues
In the dark past buried, is infinite.([12])

CHRISTIAN.

Granted then, by the power of evolution,

(12.) It is natural enough to trace man backwards nearly to the
first parents, by exploration of the different ages, such as brass, iron,
stone and bone The earth was never without these ages, and probably never will be. Civilization has never existed without barbarism;
and what is singular is, we find the ancient civilizations always on
the point of extinguishing what seem the remnants of barbarism.
The iron and bone ages have not yet left. There are regions of
barbarism unexplored. There are races that cannot work metals.
The Esquimaux may use the metalic fish-hook, if taken to him, but
he cannot make it. And further, the old seats of barbarism have
been the theatres of more ancient civilizations, whose traces are sometimes discernable. See the very instructive lecture of Alfred Russell
Wallace, F.R.S., on this subject, before the British Association for
Advancement of Science, at Glasgow, 1876.

For various species from the lower forms
To spring, the eternity of organic life
Has a fairer look. The earth yet may meet,
What in the distant past it must have seen,
Great climatic changes, hostile to life;
Still, it would be rash to say that all life,
By such changes, could ever be, on land,
Or in the deep destroyed. If eternal
We find the earth, then dates life upon it,
With God's existence.

PHILOSOPHER.

Climatic and other changes on earth
Must oft occur, which would the forms of life
Destroy or change. These changes simply mark
The limits of variation of plants,
And animals. Therefore, if by such change,
Some species disappear, and other forms,
With perfect adaptation to such change,
Are introduced, the process, all must see,
Or transmutation or evolution called,
Is not the work of chance, nor hap of fate,
Nor coaxed out result of elemental force,
But simply is, what Nature is herself,
An exercise of power divine. Such power,
Although it works by laws, may reproduce,

And seem creative, while not so in fact.([13])

CHRISTIAN.

Why not the earth its life? Each element,
(And there are not more than four score in all,)

(13.) It is a well-established law of Nature for "like to produce like." Change of varieties by amalgamation will in time become effaced—the foreign element being eliminated by the action of this law. It is a powerful, controlling law, that never ceases action. The action of this law would seem to be a veto on evolution. Species cannot change by any known law of Nature. But variations causing varieties do occur by force of law; and all those changes forced on individuals by circumstances in spite of law are mere perturbations occasioned by disturbances, never exceeding certain limits. The system of varieties to which all species are subject, is by virtue of the laws of Divine Wisdom, and necessary to promote the extension and preservation of life. If species could be formed out of varieties by mere circumstances, the system of faunas would be subverted; the occupants of one overwhelmed by those of another; and there would be no order nor repose in animated nature. Nobody objects to evolution as the work of Divine Wisdom; but, as the result of circumstances, the doctrine really seems pretty tough.

If species are eternal, there is no such thing as evolution. Every variation will fall short of the creation of a new species. Species must be eternal. There are no extinct species. Those claimed to be extinct must be mere varieties, and may be restored by generation under the proper circumstances. The mammoth was a species of elephant, and may be restored; the megatherium, dinotherium and the like, may yet be in existence in Africa, or other unexplored parts of the earth. There are also continents at each of the poles,

Has a vital principle of its own :
And each, in its normal state, is a gas.
If earth be resolved to gas, the gases,
What e'er their relative gravities, must,
In space, by well-known laws, in social group
Become diffused. A reign of death, indeed !
Life, conveyed by electric spark, or ray
Of light, may produce the material forms
In which we find them. If we say this earth,
Touched by the vital spark, from gaseous state
Came forth, with discharge of imprisoned heat,
We must expect, with all organic forms,
That it has a day of dissolution.

---

surrounded by polar oceans, whose flora and fauna are unknown,
as the Teutonic family came out from the northern polar Continent
too long ago to retain any recollection of their ancient Gladheim !
But the fossils of those realms are found in different parts of the earth,
as those of supposed extinct species. But, if thorough explorations
should show that some species are extinct, the fact will indicate that
they had a beginning. The only possible method of giving birth to
species is evolution. The species, if they ever had an origin, were
produced by the laws of Divine Wisdom, but few of whose operations and processes are known to us. The laws by which all things
are governed are eternal, and there never are any new processes under
them. If species are brought into existence, it is not probably by
transmutations from one to another. Each one is an original, and
the production of millions of germs is as easy as the production of
one.

PHILOSOPHER.

Earth die like man! The vital spark to leave,
And all resolved to gas! Is life distinct
From earth, with moulding power; or did the earth
Develope life?
      The embryo, or the germ,
A mere cell, seen by microscopic eye,
By laws of the mysterious force of life,
Assimilates the organic elements
Supplied by Nature's exhaustless stores.
The vital spark! 'tis not galvanic,
Generated in the organic cells;
But a force that, in those cells, forever
Has held its reign; reigning through organs
Cerebral, indued with the power of thought
Which conscious is, through the voluntary,
But not, when through the involuntary,
System, acting.(14) It never sleeps, and is,

(14.) Physilogists divide the nerves into two systems—the voluntary and involuntary. But, nevertheless, all muscular action induced by the nervous centres or brain substance employ thought. Those motions guided solely by the intelligence which is acquired by experience, must command attention and be conscious. There is a large class of motions which are likewise caused by thought, but are such as are unconsciously produced. There can be no muscular action arising from the motion of the brain, or nerve centre, not caused by thought, as the brain has no other way of controlling

A controlling power, and the providence
Of living things. This providence is shown
At every stage of being. The wondrous force
Called *vis medicatrix;* the instincts

muscular action; but the unconscious action is secured on a surer
foundation than human reason, and goes on independent of experience. If all the motions of the human system were to be caused by
the will, few men would be wise enough to keep their own machines
a-running.

All the principal motions and actions on which the preservation
and continuance of life depend are guided by unerring, unconscious thought. Action without consciousness needs no rest.
Reason is the exercise of higher faculties, and has charge of con-
ventional relations and duties, and is dependant purely on such
light as the individual may enjoy.

There are other motions which may be called semi-rational; they
result from the blended action of conscions and unconscious thought,
giving birth to instinct pure, and instinctive intellectual action. By
these the bird builds her nest, the bee her comb, the spider his web,
and fame and fortunes are acquired amongst men.

By saying that action is caused by thought, it must be understood
that it is not always directly so. Thought acts in two spheres.
First, on the organs of motion directly. This is unconscious.
Second, thought induces another set of motions by acting on the
will. This is always conscious.

An animal is a microcosm. From its constitution we ought to be
able to form an idea of the big world. And thus reasoning we
should conclude that all the material changes and motions of the
physical universe are produced by thought, but not thought involv-
ing the exercise of will. This work is incessant, never began, and
will never end. Its operations we call providence, and providence

That never err: the natural passions
Which, at every point of the being's life,
Promote the success of a destined end;
And then the plan or scheme, as if designed
By wisdom divine, by which the species
Are from extinction saved; this perfect round
Of well planned means to ends, invests this life,
This ruling force of the organic being,
With something more than mere material powers. (15)

it is in real fact, as all things are dependant on it, and without its constant care life would cease at once. But when we have settled this point, the book must not be closed. The same lesson teaches that there is in nature, outside of our minute microcosms, also a domain of thought and will. We simply know there is Divine Wisdom; the theatre of his voluntary rule we have not discovered.

(15.) Prof. O. C. MARSH, in an address before the American Association for the Advancement of Science, at Nashville, on the Ancient Life in America, used the following words:—"Light, heat, electricity and magnetism, chemical affinity and motion, are now considered different forms of the same force; and the opinion is rapidly gaining ground that life or vital force is only another phase of the same power."

This is so severe a criticism of the view taken in the poem, and by a so promising young naturalist and philosopher, that a notice of it cannot be omitted.

It seems to me singular that life should be likened to electricity or magnetism. These are the properties or occupants of inorganic bodies. But life is the master of organic forms. Has magnetism or

But not alone within the tissue cell
Works this magic force; but, likewise, without,

heat sensation or thought ? Life, of course, sees, hears, feels and
thinks; but it has organs for these purposes.

The body is a system of organs. Every organ has its sphere of
action. There is no action—mental, moral or physical—that is not
caused by the motion of an organ. This assemblage of organs con-
stitutes the being, and life abides in all. There is something that
keeps these organs in action; what is it ? If mental organs, the
active principle thinks. Many motions of these organs which
involve thought may take place without consciousness.

The evolutionist discovers the origin of life in a microscopic germ.
It is an invisible speck of matter animated. There it is, *vita et pre-
terea nihil.* This monarch or ruler of that germ, being active and
ambitious, commences a series of performances which should excite
the admiration of the evolutionist. It wields that germ-form along
the pathway of its simple existence, till after the lapse of a few ages,
aspiring desires evolve from its bosom a higher type of being—
perhaps an insect. No matter what. This new birth is, at first,
simply a body, nondescript as to form, invested with life. In the
course of ages, by aid or influence of circumstances called natural
selection, it acquires a set of organs that fit it for its career in its
new fauna; and by inheritance its descendants take these organs.
It should not be said, in the tones of cold philosophy, that these
acquirements, which the vital spirit has obtained through so long
and patient suffering, step in and rule the creature, to the exclusion
of the spirit itself. By these acquisitions of organs and senses, life
has merely enlarged the sphere of its powers, and added to its
capacity. But, we see, this spirit is not yet content. Insect life
may be enjoyable for a few millions of ages, when this reigning
principle will begin to feel the longings for something higher. The

Its powers are sometimes felt, and are the cause
Of what have been thought interpositions

longings may spring from the insect heart spontaneously, or be
inspired by outside influences—at all events originate or are felt in
the insect bosom. All have known, since the days of Jacob, how
impressions on the mind affect the offspring. Another evolution is
witnessed, and a being with still higher powers comes into existence.
Perhaps a bird; no matter what; say a snipe. Now, this time the
new form must be formless, as evolution does not tolerate perfect
originals. The necessity must exist before the organs. The long
legs are forced out by life's severe struggle, and are the product of
ages. The long bill, necessary for the source of its food, comes the
same way. In going through these changes that pushing principle
which we call life must experience long periods of poor living before
organs necessary to its comfort are developed. But, after all these
transitions, with all the acquisitions of ages of struggle, shall we
recognize the accomplishments as the being, and ignore the wonderful spirit that has struggled so long for their attainment ? All the
organs and qualities of the being are the acquisitions of life, which
was once but an invisible germ, almost without sense. When we
see that this simple principle may by a long series of evolutions in
the end become a philosopher, with organs of the highest order, I
think we should learn to draw a distinction between it and the
principles that govern inorganic matter.

This poem, of course, recognizes the possibility of evolution as
a phenomenon through the laws of Divine Wisdom. Life, it claims,
is as old as God and Nature, and never existed out of organic forms.
That in the changes incidental to the revolutions in Nature a reproduction of species may become necessary, and the emergencies of
all natural necessities are provided for by Divine Laws There never
was an origin of species, nor a variation that was not a repetition.

From the unseen world.([16]) Its sphere is not known.
We simply understand, it is the force
That controls existence from end to end,
Seeing all its objects attained in full.
Man is thought the wisest of breathing things:
But without control of this higher power,
His existence would find an end at once.([17])

Mere evolutionists bring species into existence in an imperfect
state, all their organs to be acquired by natural selection; but this
poem allows no origin of species, but a possible restoration of lost
ones, produced in full perfection at once, in accordance with
Divine Laws.

(16.) The human race is an unit through organic connection. The
separation into individuals is not a total disconnection. All exist in
a medium through which action and reaction are inevitable. Life
motions, of course, are through the nervous system. It is the nerve
system of the human race that is common. Certain ganglia are the
points of organic separation; but from and through these points
individuals are kept in union by the out-reaching and all-pervading
power of the vital spirit. Providences frequently come through
unexpected kindnesses of fellow beings, whom, we are apt to think,
the Divine One has moved; and the Divine One may be within the
circuit of vital action. Nature's language says yes; and in prayer
for relief in distress, it can do no harm to appeal to the Source of all
Power, as well as to our fellow creatures.

(17.) Prof. Tyndall, in his address before the Midland Institute,
at Birmingham, Oct. 1, 1877, expressed his entire consent to the
adoption, by his hearers, of an ideal soul. Said he:—"If you are

When the individual's career is o'er,
And every object of his life obtained,
What then remains, and what its destined end?
Many look with dread on dissolution,

content to make your 'soul' a poetic rendering of a phenomenon which refuses the yoke of ordinary mechanical laws, I, for one, would not object to this exercise of ideality." (See London *Times*, Oct. 2.) But this imaginary entity or spirit the learned lecturer proceeds to load with chains, entirely destroying the freedom of its will. At first thought this servitude of the will to the laws of being might seem a grievance, and was undoubtedly thought by the learned professor to detract from its dignity. If science would only turn her eye from things of darkness to things of light, the view would be changed. It is a grand fact that Nature, whose sphere is the universe, is a system of laws. At once it is perceived that the beings of earth, unsupported and uncontrolled by laws, and endowed with free will, could not continue in existence. In most of their motions and actions they are coerced by involuntary faculties; and so far as volition has sway, an irresistible motive governs. The being, even in the limits of his selective privileges, is ruled by his judgment of what is best. This, we have reason to infer, is not a condition peculiar to earth or earthly life, but must be the mode of existence of all beings, in all spheres, embracing the highest of all. The Divine Being of Revelation cannot act contrary to Divine Laws. And the nearest approach by society on earth, to celestial happiness, is when men are the most religiously and perfectly governed by Divine Laws. Of course, but for the imperfection of the intellect which rules the voluntary actions, all would choose the best, as the apparent best is the controlling motive. Obedience to Nature's laws is the rule of life, and is, undoubtedly everywhere, on earth and in heaven, the only road to peace and felicity. Then, granting that

As though life could ever become extinct;
But such forebodings are but borrowed fears.
Death is a phantom; not a real foe.
All creeping, breathing things, are here fixed fast,

there is no free will, as Mr. Tyndall claims, does the admission of
this fact aid the inference he struggles to raise, that man is a machine ?
—that his consciousness or power of thought is the mere ticking of
a clock ?—a motion caused by the vibration of molecules ? Whether
or not intellectual action is caused by anything else than molecular
motion the learned professor says he does not know. Science traces the
sensory nerves to their centre, and sees the effects of commands sent
out to the muscles through the motor nerves; but, really, it would
be unscientific to infer that there was a commander within that
centre ! And still, there lingered in the professor's breast a feeling
that there exists something like a soul, and he permits us to enjoy
one in idea; and the great German Prof. Haeckel, in a recent address
at the Munich meeting of the German Association, gave the world an
absolute article, which he called "the plastidule soul." Now, with Mr.
Tyndall, let us note the effects of sensation. The child, admiring
the blazing fire, puts his hand in it. The burn sends its sensation
by sensor nerves to the nervous centre in the brain, where the organ
of motion centres. From this organ, through the motor nerves, the
command goes to the muscles of the hand, to remove it from the
fire. This action of the motor organ is caused, as said by the lecturer, by molecular motion. If observation ended here, there would
be more obscurity. But it is to be noted that ever after this experience, whenever, by the vision of fire, its sensation is conveyed
through the optic nerve to that central organ of motion, the old
sensation of its burning quality is remembered, and due warning
given, with proper orders, as circumstances may require, to the
muscular members to take protective action. The sensation causes

And must here remain. Eternal wisdom,
With which we find this endless life imbued,
Has lodged the highest joy in youthful veins,
And hence permits, not death, but renewals
To everything that lives. Each germ thrown off
By the parent plant, animal, or man,
Is that parent's self—is his very being—
With all his habitudes, moral, mental
And physical, indued. Throughout his life
Such germs must be, in numbers vast, evolved,
And in them, as they grow up around him,
Does the parent, with life renewed, exist.
These emanations from the parent stock
Draw each a part of the parental life,
And are, as is seen, but transition steps
Of the paternal being, who thus goes forth,
In youth and beauty, in life's new career.
Thus renewed, the new form survives the old,

the molecular motions to strike the motor organ, the fibrous centre
which is thereby made to act; and if this fibrous centre organ did
not possess an indwelling spirit, how could that sensation be recalled
by the action of simple fibres ? This power of remembering past
sensations, coupled with the exercise of perception, judgment and
volition, might, without particular detriment to the dignity of
science, be attributed to a force which reigns in organic cells; a
communicated force, if those cells had an origin; or an eternal one,
if they are eternal. ·

Which yields to dissolution : but this fate—
Which men call death—when Nature's path is trod,
No terror has—has no pang.([18])   One by one,
As age advances, the ruling passions
And strong desires of early days, depart,
Till the mind itself is almost a blank.
Soon memory ceases to recall ideas
Of events or forms; and the aged one,

(18.) Death to the young, or one in the vigor of his faculties, is a distressing calamity; to the aged, whose faculties have decayed, a blessed benefaction. In the last case life has been gradually passing from the parent to the child for scores of years. Amongst Nature's greatest mysteries are the laws of generation, whereby dual beings are enabled to enact the part of a creator. New beings are produced—drawn from the vitals of the parents. The offspring, for soul or body, has nothing not emanating from the progenitor, and is the progenitor renewed. This progeny is cradled in the affections of its parent, and to maturity is fed and fostered by parental love, and owes its being as much to its parent's brain and muscle as if brought up inside of his frame. The labor or force of the progenitor is all invested in his second self, till his own life is merged in the new life, when his exhausted faculties cease longer to act. The new man has really grown out of the old one, and takes his place. No matter when the judgment day comes, whether in five or fifty millions of years, we shall all be on hand. It is to be hoped that, when the dread day comes, by aid of these continual renewals, we may become prepared for its awful exercises. If we are to be judged as we shall then appear, it is all important to take particular pains with our renewed persons, as labor in the right direction for future happiness.

Fed like a babe, step by step, weaker grows,
Until naught but the withered form of man
Remains. By slow degrees, the senses all
Are seen to fail; the eyes no longer see,
Nor longer hear the ears, nor any taste
In the palate left. But ere the lingering
Vital spark shall go out, all sense and thought
Will long have ceased. The change is then complete;
And the monumental stone simply marks
One stage of man's existence.[19]

(19.) In treating of evolution, all reasoning must be from phenomena of nature, and without reference to Scriptures. But few evolutionists fail to see in the works of Nature the existence of Divine Wisdom. The idea of a creation is combatted, and evolution brought in to supply its place. If evolution be an established law of Divine Wisdom, its operations are really creative; but if the result of circumstances, then the world came by chance.

Of course no evolutionist refers to a future existence of the soul, nor does the poem speak of any end except that accounted for by natural laws. Still, Nature is not supposed to be without evidences that the vital principle may survive the body. Dr. Darwin, the grandfather of the great naturalist, nearly a hundred years ago pronounced the belief of such survival not unphilosophical. In his *Zoonomia*, speaking of the spirit of animation, he says :—" This immaterial agent is supposed to exist in or with matter; but to be quite distinct from it, and to be equally capable of existence, after the matter now possessing it, is decomposed. Nor is this theory ill-supported by analogy, since heat, electricity and magnetism can be given and taken from a piece of iron, and must therefore exist,

CHRISTIAN.

With your philosophy, I find no fault.
You grant me God; and, in organic life,
You find a providence; and races new,
Evolved from antecedent races, which,
In one sense, you say, is called creation.
That God is eternal, I will concede.

whether separated from the metal or combined with it. From a
purity of reasoning, the spirit of animation would appear to be
capable of existing separately from the body, or with it."

But whether this spirit of animation or life can be imparted to new
bodies except by the natural course of generation, or not, is a subject about which minds will differ. However transmitted from old
to new forms of being, whether physically or spiritually, it is clear
that it must be a transmission of the vital principle itself, destitute
of any recollection of its prior existence. The impressions the body
receives must perish with it—the spirit alone surviving. Of course,
it is natural for men to wish to take with them into heaven the
splendid stock of wisdom acquired by them during life ! The loves,
the hates, the malice, the prejudices, the errors, the knowledge of
life's struggles, the pride, the presumption, the exceeding self right-
eousness—in fact, all the petty recollections, of sometimes a pretty
mean existence—are hugged by the dying, but repentant sinner, as
a precious stock, calculated to fit him for associating at once with
angels of light ! If the spirit of man enters the portals of light, as
we fondly expect, it must, in the language of the Scriptures, be a
new birth. The offspring of births take the parents' existences,
minus recollections of their experiences. This is a wise law, otherwise socially we should be badly mixed in the other world.

If earth began, 'twas not the beginning
Of God, nor of his world; but must have been
One stage, only, in an infinite round
Of transformations, which never began,
And, hence, will never end. God's magnitude
Will not be compressed to the dimensions
Of man's philosophy. We must enlarge
To take him in, or our structures fall
At his approach. Man is a finite being,
And must build his philosophic systems
With movable timbers, because often
He will have to alter and enlarge them.
I think well of your speculations, for
It is the true sphere of philosophy
To find out God, and not ignore him.